RIVER

A series of Riverine Small Books

by Sylvia M. Haslam and Tina Bone

BOOK 6

(*A River Friend Series Reference Book*)

An Introduction to the
WATER FRAMEWORK DIRECTIVE

Sixth book to be published:

An Introduction to the
WATER FRAMEWORK DIRECTIVE
(A River Friend Series Reference Book)

A Book in a series of Riverine
publications by

Sylvia M. Haslam and Tina Bone

(Each book is about a different subject so the
series can be read in any order)

*Written and Edited by Sylvia Haslam and
Tina Bone. Illustrated by Tina Bone
(unless otherwise stated)*

RFS6: PAPERBACK **60** pp.
ISBN No. 978 1 9162096 3 3
28 Illustrations

Published by: Tina Bone UK
First edition: **December 2020**

www.riverfriend.tinasfineart.uk
Email: ourbooks@tinasfineart.uk

HOW TO USE THIS BOOK (RFS6)

1. Read pages 1–9, which describe what and why there is a **Directive** about water.

2. Skim through the **Articles** on pages 10–17, reading their headings.

3. Using the **Contents** list below and pages 10–17, identify your areas of interest. (Note that *Definitions* are given on pages 10–11).

4. Study the—now short—text.

5. This book, like the **Prologue** to the River Friend series (*Plant Identification and Glossary of Terms*)—but unlike others in the series—is more for reference than for study. We hope it may help you understand a little bit more about this very useful Directive, which has become Law in EU member countries, and this still includes Great Britain until repealed.

6. A **screen-view version** of this little book (RFS6) is available to download free from the River Friend Website:

   http://riverfriend.tinasfineart.uk/resources/

CONTENTS

INTRODUCTION TO THE SERIES

Rivers are vital. They bring freshwater to the land, on which all its life depends. They are beautiful and fascinating, making up both the typical British countryside and many of its most spectacular views. If they vanished, what hardship and outrage there would be! Yet, slowly, slowly, they are vanishing, the larger stream becomes smaller, the tiny brook becomes a ditch and dries, and is filled in— the small ditches get polluted and dug out, become dull, and vanish from sight and consciousness. How can we save our rivers and riverscapes? How can we raise awareness on this slow, almost invisible loss?

We believe that this series of handy, small books, suitable for readers from teenage upwards, will help to raise awareness. Individually, each book tells a story on a particular riverine and riparian environment. Collectively, the series will inform, in a simple and effective manner, the extraordinary value of freshwater and its plants.

The Authors realised that there was a huge gap in the literature. There are many publications for scientists, for pond-dippers, birders and anglers, but "easy-read" books focussing on the river itself, and the vegetation belonging to it and creating the habitat for all else: we could find none!

For explanations regarding British freshwater plants, terminology mentioned throughout the series, and Picture Guide and reference section for further reading, see the book entitled *A PROLOGUE TO THE SERIES: Plant identification and Glossary of Terms* (also available to view in pdf format free on-line at http://riverfriend.tinasfineart.uk/resources/

Other titles in the Series are listed on the last page of this book and on the River Friend Website:
http://www.riverfriend.tinasfineart.uk

An Introduction to the
WATER FRAMEWORK DIRECTIVE

INTRODUCTION

WHY THE BOOK?

RAIN-TO-RIVER-TO-SEA: everyone should be able to aid waters, to make some water, somewhere, less polluted, less disturbed, less upset: and do so without nuisance or expense to oneself. Fewer of us can do big things needing planning and expenses but little things, done by the million, add up. "Many a mickle makes a muckle" as the old saying goes.

This is one of several Books in a series written by Sylvia Haslam and Tina Bone which describes, in uncomplicated terms, the informed efforts of some of the world's best water scientists to convert water of death or some life to water of good, or at least better, life in the streams, rivers, lakes and groundwaters of Europe.

The "River Friend" series of books introduces research material to a general readership. Ecological material is very seldom difficult in itself. It appears so because it is usually written from one research scientist to another in a research journal. But this is not necessary. Ecological principles are usually obvious once

1

pointed out. (A common example is storm run-off from roads, running into rivers. Think of a motorway. Think of all the dirt from cars and lorries, bits of tyres, bits of oil and petrol, quite apart from everything from sandwiches to a milk lorry turned over. Is it not obvious that the storm water is polluted? So will harm the river? Yet how many readers have thought about it?)

Gone are the days when people went out with buckets and came back with the nice clean water they needed. People are just too many. When regulation fails, famine and disease strike. Water policy must be there to keep the population healthy, and transport and industry working well.

The European Union, of course, sticks its fingers into water policy—indeed not just its fingers! Water is far too important for that! However, the policy document, The Water Framework Directive, is not all written in an easy style. For instance:

(21) The Community and Member States are party to various international agreements containing important obligations on the protection of marine waters from pollution, in particular the Convention on the Protection of the Marine Environment of the Baltic Sea Area, signed in Helsinki on 9 April 1992 and approved by Council Decision 94/157/EC, the Convention for the Protection of the Marine Environment of the North-East Atlantic, signed in Paris on 22 September 1992 and approved by Council Decision 98/249/EC, and the convention for the Protection of the Mediterranean Sea Against Pollution, signed in Barcelona on 16 February 1976 and approved by Council Decision 77/585/EEC, and its Protocol for the Protection of the Mediterranean Sea Against Pollution from Land-Based Sources, signed in Athens on 17 May 1980 and approved by Council Decision 83/101/EEC. This Directive is to make a contribution towards enabling the Community and Member States to meet those obligations.

The above extract, we think, should show why we have written this Book.

THE WATER FRAMEWORK DIRECTIVE—THIS BEING THE USUAL SHORT FORM OF THE DOCUMENT PROPERLY KNOWN AS:

EU DIRECTIVE 2000/60/EC OF THE EUROPEAN PARLIAMENT AND OF THE COUNCIL OF 23 OCTOBER 2000, ESTABLISHING A FRAMEWORK FOR COMMUNITY ACTION IN THE FIELD OF WATER POLICY.

Amended by:

Decision No. 2455/2001/EC of the European Parliament and of the Council of 20 November 2001

Directive 2008/32/EC of the European Parliament and of the Council of 11 March 2008

Directive 2008/105/EC of the European Parliament and of the Council of 16 December 2008

Directive 2009/31/EC of the European Parliament and of the Council of 23 April 2009

Directive 2016/39/EU of the European Parliament and of the Council of 12 August 2013

Council Directive 2013/64/EU of 17 December 2013

Directive 2006/118/EC of the European Parliament and of the Council of 12 December 2006 on the protection of groundwater against pollution and deterioration—and no doubt by future Directives and Decisions also.

UK LAW: STATUTORY INSTRUMENTS 2017 No. 407 WATER RESOURCES, ENGLAND AND WALES. THE WATER ENVIRONMENT (WATER FRAMEWORK DIRECTIVE) (ENGLAND AND WALES) REGULATIONS 2017.

Made 15th March 2017
Laid before Parliament 16th March 2017

Laid before the National Assembly for Wales 16th March 2017
In force from 10th April 2017

"The Secretary of State and the Welsh Ministers, acting respectively in relation to river basin districts that are wholly in England and river basin districts that are wholly in Wales, and jointly in relation to river basin districts that are partly in England and partly in Wales, make these Regulations in exercise of the powers conferred by section 2(2) of the European Communities Act 1972(1)."

3

For all

The EU Water Framework Directive (WFD) **applies to all of us**. It is the best shot by the EU to make our brooks, rivers, lakes and groundwater better: to give us cleaner water, safer water resources, and less damaged water patterns and processes. And a very good shot it is too! In the authors' opinion it is the best and most comprehensive work describing good and high status that there has ever been.

If you have never read the WFD, do take a few minutes to read through this book quickly. Find out why we need the Directive (easy—because no country was or is doing as well as it should, and could, be doing). What does the Directive recommend and require? Each nation (country, Member State) was required to transpose the Directive into its own national laws. As with much EU policy, the actions did not have to be done at once. It was to be brought in over time. And as with many laws, national or EU, not all parts are given the same weight or urgency by different groups of its audience. But all are in fact Law. National Governments can add to the Directive. Denmark's excellent decree of buffer strips 2m wide beside each and every channel carrying flowing water, whether tiny rills or large rivers, is one such, and pre-dates the WFD. But they cannot impose this on their neighbours without another Directive, and they cannot remove articles or actions from the script.

Because the WFD applies to so much of life—water supply, quality and quantity, hence riverscape, hence pollution, hence river life, hence town and country planning, hence energy resources, hence those who are implementing the Directive officially must of course study the original. This book is insufficiently detailed. It is enough for most members and volunteers of, for example, the Wildlife Trust, but those who are Trust staff should read the whole Directive.

(Comments by the authors have been put in *blue italic type* in square brackets; otherwise, after this beginning, sentiments expressed are from the Directive itself up to the concluding "Afterword".)

PREAMBLE

WHATEVER IS A DIRECTIVE?

Parliament makes laws. These laws are Acts of Parliament, and Parliament expects all those who are addressed, to obey them (for example, buy car insurance, pay your taxes). In the same way, the European Parliament makes laws. These are called "Directives", and are for the whole European Union. (Directives go to the Parliaments of each member country, to be made national laws.)

BUT WHAT IS IT?

The European Parliament and the Council of Ministers, that is, the top, most weighty part of the European Union, say that:

> *"WATER IS NOT A COMMERCIAL PRODUCT LIKE ANY OTHER BUT, RATHER, A HERITAGE WHICH MUST BE PROTECTED, DEFENDED AND TREATED AS SUCH [i.e. as Heritage]."*

The remaining nearly 100 pages of the Directive set out how these high bodies (at the behest of their scientists) consider the best ways to protect and defend water, and value it as Heritage.

SO WHY IS IT WANTED?

Just by existing, each of us makes rivers worse. Each person drinks, so uses up water, though some re-appears as polluted waste water. This may go quickly or slowly to the sea, and be slow or very slow to evaporate back into the sky as clouds and eventually rain. Each plant and animal also needs water, and again only some re-appears. Much is lost to the air (from where it eventually re-appears as rain, but probably not in the same place). Much, again, goes to the sea.

Each of us requires shelter, warmth, goods and chattels: things made by people, normally involving water, for the making, the transport, the selling—even sales staff eat and drink!

Our population is growing. In Britain in the 1940s, 40 million was the accepted figure. Now it is nearer 68 million. Economies of scale do not apply, two people drink twice as much as one, and produce twice as much waste. Ten people drink ten times as much as one, and so on. Ever-increasing standards (so far) mean more water consumption. The water is demanded, cleaner, no

5

longer just whatever is in the nearest stream or pond. More goods and chattels are demanded. Have you ever thought how much more water is used in making and transporting furniture for an 8-room house than for an ancient one-room cottage? For a full supply of cutlery and crockery (to name, but one thing) than for only a knife each—and most "plates" being slices of bread?

No wonder water is running out!

No wonder the **Water Framework Directive** is needed!

And, unfortunately, no wonder many people want to ignore and bypass the Directive, in just the same way many people would like to bypass paying Income Tax or not driving whilst drunk, to list two other lawful measures established for the common good.

Something should be done, as waters in the European Community are under increasing pressure. Demand for much water of good quality for all purposes is continuing to grow. It is for the European Union, including all of us in Britain, to do it: to preserve, protect and improve environmental quality and ensure prudent and rational use of natural resources. Programmes should be based on the precautionary principle that, if there is no evidence something is safe, it should not be done. The normal procedure seems to be the reverse; to do it unless it is known to be dangerous—which causes too many disasters. Environmental damage should be prevented and if possible be put right at its source, and where possible the one doing the damage should pay—for example, the polluter for pollution removal.

- Water supply is a service for the general good.
- Good water quality helps to provide the drinking water for the population.
- Policy needs a "framework" (a frame within which to work) which is transparent, effective and coherent.
- The Community (of European nations) should provide principles and a framework for action. The Directive should integrate these and further develop the overall principles and guidelines.
- Whilst the primary purpose of the Directive is water quality, doing this requires ensuring water quantity also, so both quality and quantity are in fact covered. (Since groundwater may influence water flowing or standing beyond or above it, it must also be considered.)
- For this, definitions of good quality water (and where relevant, good quantities of water) need to be established equally throughout the Community.
- Deterioration of water status should be prevented.

- **To achieve or, if already Good, maintain status, actions must usually be taken. These actions are known as "Programmes of Measures".**

- The final aim is to get rid of priority hazardous substances, for example benzene, and reduce the levels of natural damaging substances such as nitrates to those near what would be background levels.

- Surface and groundwaters are, in principal, renewable natural resources, but there is a delay, particularly for groundwater, for it to gather and renew. The time lag must be considered when establishing measures for achieving good status, and for reversing any sustained increase in pollutants in groundwater. Therefore there are delays in implementing parts of the Directive. *[Apart from all the other causes of delay and emission!]*

- There are water bodies which are in such a bad state that it would be unreasonable to turn them into Good status. Here aims can be lowered, though all practicable steps should be taken to prevent further deterioration.

- There may be unforeseeable or exceptional circumstances for exemptions from these requirements, for example, floods and drought, overriding public interest, new modifications or alterations. However, all practicable steps must be taken to lessen adverse impacts.

- The aim of achieving good water status should be pursued for each river basin (catchment), so that ecology, hydrology and hydrogeology are co-ordinated for each. Similarly, qualitative and quantitative aspects should be integrated (for example abstracting more water may nullify the good effects of putting in fish passes and removing more phosphate).

- No surface water means no fish! All characteristics of a river basin and its human activity need to be looked at, and so must the economics of water use. These should be done in a comparable way throughout the European Community area.

It is all very well to talk about EU being Law—but in June 2016 a British Referendum (by a small majority) voted to leave the Union rather than remain in it. So what now? What is quite sure is that until Parliament has acted, existing EU Law remains as National Law. All the EU Acts and Statutory Instruments were passed by Parliament, and only a Parliament can undo what a Parliament has done.

Given that there are thousands of EU Acts, there cannot in practise be a blanket "undoing" of all of them, just like that. For a start, many—like the WFD—are clearly good for Britain, and should be kept, in whole or in part. How long will it take to examine all, and decide which are really good for us, and is this in whole or in part? *[Years is our guess.]* Next, in order to keep in the Common Market, Britain would have to accept (not reject in Parliament) all the Directives governing the Single Market: good, bad or indifferent for Britain.

While this is obviously a nuisance, if not worse, it is also sensible and right. If buying an EU television, car or cushion, the British undoubtedly want it to be of good quality (or, quality appropriate to the price) and fit for purpose. That is, goods sold must meet EU standards and with 28 countries involved, and thousands of rules, it is unfortunately not practical for Britain to say it will keep this, but not that—would we want to change the size of batteries, or allow diseased foreign cattle to be imported? Obviously much negotiation is needed—and needs continually to be carried out.

This also means no drastic changes will be done in a hurry. The WFD remains in force in the form presented here, until the British Parliament decrees otherwise. Readers in 2020 and beyond should ask the current position from their local Environment Agency or Rivers organization.

This book is not an EU document—official or unofficial. It is a summary of a very important Directive. It follows the Directive and includes extracts from it, but its main purpose is, as described above, to introduce readers to something pervasive in its scope, and valuable in its holistic approach. It is NOT a manual, and before taking any action to implement the Directive, readers MUST consult the original, and follow the instructions and directions in that (together with such addenda as their own Governments have decided upon).

But can just anyone write about a Directive? The standard for copyright is that up to 10 per cent can be quoted. However, the EU thinks Directives should be available to all. *[How wise!]* Hence there is a European Commission Decision of 12 December 2011 on the re-use (copying, etc.) of Commission Documents (2011/833/EU) confirmed by a Legal Notice of 2015. Re-use of the EUR data for commercial or (as in this book) non-commercial purposes is authorized provided the source is acknowledged. (The only exceptions are, for example, unpublished personal research data.)

- Water bodies used for the abstraction of drinking water should be identified, and principles laid down for its control to ensure environmental sustainability.
- Accidental pollutions affecting waters should be prevented so that measures concerning these should be included in the Directive: controlling pollution at source (emission limits) and setting quality standards).
- Pollution by priority hazardous substances (see Annex 2) should be phased out and stopped. The control measures should identify these substances, consider the precautionary principles, the adverse effects and risks, all significant sources, as well as the cost-effectiveness of the controls. Other pollutants influencing water bodies should be progressively reduced, similarly.

8

- To get the general public interested and involved, proper information and reports on progress are necessary, particularly when final decisions are adopted. *[Wow! The general public to hear about the Directive, and what is happening? Whatever next?]*
- Each year the Commission (of the European Communities) should present a plan for any new initiative.
- A thorough understanding and consistent application of the criteria defining river basin districts and evaluation of water status are needed, so guidelines may be needed on how to apply these criteria. (It follows that there should be penalties for disobedience, which should be effective, proportionate and dissuasive.)

FOR ALL THESE REASONS THE EU HAS ADOPTED THIS FRAMEWORK DIRECTIVE FOR WATER STATUS, WHICH HAS 26 ARTICLES, 11 ANNEXES (AND IS ADDED TO BY LATER DIRECTIVES AND DECISIONS).

Stoneywood, Aberdeen, Scotland

River Ver, behind Park Street, St Albans, Hertfordshire.

THE ARTICLES

ARTICLE 1. PURPOSE *[The Purpose of the Directive]*

(a) To prevent further deterioration, to protect and to enhance aquatic ecosystems, and, for water, the lands and wetlands directly depending on them.
(b) Promote sustainable water use, based on long-term protection of available water resources.
(c) Increase the protection and improvement of the aquatic environment, through reduction, phasing out or stopping pollution to both surface and groundwater.
(d) Help to mitigate the effects of floods and droughts.

By doing this, the Directive helps to provide sufficient good-quality water, to reduce pollution, and to protect water bodies, both on the surface and underground.

ARTICLE 2. DEFINITIONS

Wisely, the Commission wishes to be clear about the meaning of its statements. A selection of the terms defined is:

Inland water means all standing, or flowing water, on the land surface, and below ground. Water there and on the landward side of the edge of territorial waters.

River means inland water flowing mostly on the land surface (though it may partly flow underground).

Lake means a body of standing inland surface water.

Artificial water body means a body of surface water created by human activity.

Heavily modified water body means a surface water body which has been substantially modified by human activity.

Body of surface water means a separate and significant surface water, such as a whole or part of a river.

Aquifer means subsurface rock of enough porosity and permeability to allow a good flow of groundwater, or abstraction of significant quantities of groundwater.

River basin (catchment) means the area of land from which all surface run-off flows into the sea at a single river mouth.

Surface water status means whichever is the worse status of the ecological and chemical status of a surface water body. [Note that these two—ecological and chemical status—are given equal importance.]

Good ground surface water status is when both ecological and chemical status are "Good", or better than that.

Groundwater status means whichever is the worse of the quantitative and chemical status of a groundwater body.

Good groundwater status is when both quantitative and chemical status are "Good", or better than that.

Ecological status is the quality of the aquatic ecosystems in its structure and function. This is described in Annex 5.

Good ecological status is what it says.

Good ecological potential means that a heavily modified or artificial water body could, according to the criteria in Annex 5, improve.

Good surface water chemical status is chemical status classified as "Good" under Articles 4 and 16, Annex 9, etc.

Good groundwater chemical status is chemical status classed as "Good" under Annex 5.

Available groundwater resource is the difference between the extra incoming water and the out-flowing water of the aquifer. If more than this amount is taken and used, the ecological status of the water body and its dependent land communities deteriorate. So the available resource is that which can be abstracted sustainably.

Hazardous substances are those listed as toxic, persistent and are able to bio-accumulate *[that is, be stored up in plants or animals]* in other substances of equal concern.

Priority substances are those dangerous pollutants defined and listed in Annexes 2, 6 and 10. Included here are the priority hazardous substances, for which special measures are laid down.

Pollution means the introduction, direct or indirect, by human activity, of substances or heat into the air, water or land which may be *[note, not are proven to be]* harmful to human health, to the quality of aquatic ecosystems, or to land ecosystems dependant on the water ones. Such pollutants may also damage material property, or impair or interfere with amenities or other proper uses of the environment.

[This is one of many definitions of pollution. It is a truly excellent one, particularly the first part, and is worth memorising!]

Environmental quality standard means the pollutant concentration which should not be exceeded for the protection of human health and the environment.

Water Services means those services providing water for households, public institutions or any economic activity. These involve (a) collection of surface or groundwater by abstraction, impoundment, storage, treatment and distribution and (b) wastewater collection and treatment where this later discharges into surface water.

Water use means water services and other activities (Article 5, Annex 2) significantly altering water status.

Emission limit values mean the levels of substances which may not be exceeded (see, in particular, Article 16).

Emission controls are the controls needed to achieve the above standards.

ARTICLE 3. CO-ORDINATION OF ADMINISTRATIVE ARRANGEMENTS WITHIN RIVER BASIN (CATCHMENT) DISTRICTS

[This allows for small river basins (catchments) to be combined with others, as in W England and, more disastrously, Malta, where catchments differing in geology, landscape, vegetation, land use, etc., have been combined for convenience. Otherwise, this Article is concerned with administration, and is either needed in full, by official bodies, or not needed, as by most individual researchers.]

11

ARTICLE 4. ENVIRONMENTAL OBJECTIVES

To achieve their proper aims, member states shall:

(a) for surface waters, prevent deterioration, protect, enhance and restore natural water bodies, aid artificial and heavily modified waters, and lessen pollution. All this is specified to be done within 15 years, i.e. by 2015, raising their status to Good or better, as described above.
(b) to similarly protect, prevent, enhance, restore and aid groundwaters, and to ensure that abstraction does not exceed recharge.

When more than one objective appears to be in breach, the control measures for the most stringent breach shall apply.

Water bodies may avoid the requirements for these improvements when doing so would disrupt the wider environment: navigation (in whole or in part), recreation, water storage, water regulation, flood protection, drainage or other equally important sustainable activities. To avoid deterioration from inertia, management plans must be reviewed every 6 years!

Details are specified of when and how the objectives and their time-scale can be modified for any particular water body.

ARTICLE 5. CHARACTERISTICS OF THE RIVER BASIN DISTRICT, REVIEW OF THE ENVIRONMENTAL IMPACT OF HUMAN ACTIVITY AND ECONOMIC ANALYSIS OF WATER USE

[Here we go! More in the Annexes.]

Each river basin (catchment) shall have an:
* analysis of its characteristics
* the human activity influencing waters
* economic analysis of water use.

[This was to be done by 2004—what a hope! And was to be updated by 2013 and every six years thereafter. Really?]

ARTICLE 6. REGISTER OF PROTECTED AREAS

There shall be none.

ARTICLE 7. WATERS USED FOR THE ABSTRACTION OF DRINKING WATER

These shall be identified, and properly protected and maintained.

ARTICLE 8. MONITORING OF SURFACE WATER STATUS, GROUNDWATER STATUS AND PROTECTED AREAS

[This means there must be surveys, monitoring and programmes of actions on water status. What is in the river? How good or bad is it? Until these are known, how can status be protected or enhanced or even maintained? (Of course much has been done over decades and centuries, but not throughout the EU or in a consistent manner.)]

Programmes shall cover, for surface water:
- the volume and flow needed for ecological and chemical status and ecological potential
- the ecological and chemical status and ecological potential for groundwaters as well as surface ones
- the chemical and quantitative status

(and for protected areas, the extra specifications needed for each).

This is merely the general purpose. What is ecological status, for instance? Annexes 2 and 5 describe what is needed! The methods for analysis and monitoring have to be laid down to ensure, as said above, that there is consistency. *[Unfortunately, different laboratories with the best will in the world, and the same methods and protocols, do come to different conclusions from the same river data and samples. This is life—but it must be remembered when dealing with results. Too often data entered is treated as set in stone and false conclusions are drawn, and even worse, false measures taken.]*

ARTICLE 9. RECOVERY OF COSTS FOR WATER SERVICES

This should be done, but within the specifications in the Article. Member States may, in doing this, consider the social, environmental and economic effects of recovering money as well as the geographic and climatic conditions of each region (also see Annex 3).

ARTICLE 10. THE COMBINED APPROACH FOR POINT AND DIFFUSE SOURCES

Sources of pollution, that is. Early work and approaches tended to separate these, point sources being sewage works, factories, and suchlike, and diffuse ones, barely considered, being agricultural and road run-off and suchlike. With the increase of both population and regulation, the distinction between them is now blurred: road run-off from a little-used back lane draining into a stream every few metres is diffuse pollution, but what about the gathered run-off from many kilometres of a busy motorway? Both are road run-off, and they grade into each other.

Pollutant discharges are to be controlled *[but this is an aim, rather than something presently achievable].* There are three approaches:

(1) emission controls based on best available techniques: briefly, do not let pollutants reach the water body;
(2) limit emission values: if it is in the water, if it goes over the specified limit, action must be taken; and
(3) for diffuse pollutions, use the best environmental practises to lessen or lose this, i.e. fertilizer pollution can be reduced by lessening the amount put on, the type put on, by not putting it on before storms, by having blocks to its running off (ditches, hedges, furrows perpendicular to slope), and so on.

Earlier Directives still apply.

ARTICLE 11. PROGRAMME OF MEASURES

[Many who do not know where this phrase comes from, or what it refers to, have heard it and heard it as something bad. A friendlier phrase might have been better for public relations! Would "How to Improve", have gathered such a bad reputation?]

Anyway, the Programme of Measures are the actions to be taken to achieve the protection, enhancement and restoration of water bodies as listed in Article 4 (above), taking account of the results of the surveys, etc., required in Article 5. These are the nitty-gritty of what needs to be done for water, indeed not just for "water" but for each water body, river basin (catchment) or river basin district. Over the EU the variation in catchments, waters and their present status is immense. It follows that the range of programmes, applied to each, should be equally immense.

The minimum requirements for each river basin (catchment) are termed "Basic measures". They can be divided into:
- protection measures, to implement laws (including Article 10 and Annex 6A)
- cost-recovery measures (Article 9)
- measures for environmental objectives (Article 4)
- measures controlling abstraction (Article 7), including registers of abstractable waters
- emission controls on point source pollutant discharges, and controls of diffuse source pollution
- measures to ensure the hydro-morphological conditions fit with the ecological status or potential
- measures to prohibit direct discharges of pollutants into groundwater (except for specified exemptions. One of these is geothermal water put back into its own aquifer. Another is for limited scientific purposes. The exemptions may not be enough to hinder the achievement of the environmental objectives.)
- reduction or prohibition of the priority hazardous pollutants (e.g., pollutions due to floods can be foreseen, though their dates probably cannot, and some mitigation is possible).

Supplementary measures are needed where the basic measures cannot achieve the aims. Stricter environmental standards may be needed: or alternatively additional measures may not be practicable (e.g., to avoid earthquakes).

The Programme of Measures shall be reviewed, and if necessary updated, in 2015 and every six years thereafter. Updating shall be implemented within three years. *[Unfortunately, not even the EU's firm use of the word "shall" can make anything happen.]*

ARTICLE 12. ISSUES WHICH CANNOT BE DEALT WITH AT MEMBER STATE LEVEL

Report to European Commission, etc.

ARTICLE 13. RIVER BASIN (CATCHMENT) MANAGEMENT PLANS (RBMPS)

Each river basin (catchment) shall have its own management plan (which includes Annex 7 information). First plans published in 2009.

The Secretary of State for the Environment, Food and Rural Affairs has approved the revised RBMPs of 2015—to be reviewed every 6 years. "We have prepared them in line with ministerial guidance, they fulfil the requirements of the Water Framework Directive and contribute to the objectives of other EU directives."

ARTICLE 14. PUBLIC INFORMATION AND CONSULTATION

[Good!]

It is important that all interested parties, not just officials, are involved in producing, reviewing and updating these plans. Plans should be published and available at least three years before they are implemented. *[Very wise. Time is needed to get information widely known, and to arrange any recording, or analyses.]*

ARTICLE 15. REPORTING

Reporting and publications of analyses (Article 5) and monitoring (Article 8) shall be available within three months of their being finished. Within three years of the publication of each plan or update there shall be an interim report describing what is happening in its Programmes of Measures.

[Fortunately, it is very difficult to conceal lack of progress. Short reports and lack of evidence and detail are telling! So when little has been done, or little has been achieved, it is usually possible to detect this—provided timely reports are available. So, too, can good work be noticed and praised. All should learn, though, to detect when data have been (in the words of W.S. Gilbert) added "to provide verisimilitude to an otherwise bald and unconvincing narrative".]

ARTICLE 16. STRATEGIES AGAINST POLLUTION OF WATER

This is a complex Article, describing how to recognize and assess those substances listed as priority hazards (Annex 2). These pose risks to, or via the aquatic environment as assessed by their intrinsic hazard (aquatic eco-toxicity and human toxicity via water), by monitoring widespread environmental contamination, and by other "proven factors" (e.g., production or use volume of the substance, and use patterns).

The list of priority substances shall be reviewed at least every six years.

There shall be proposals for quality standards for "cost-effective" product and process controls for both point and diffuse sources of pollution, and emission limits. Proposals shall include reviews, updates and assessments of their effectiveness.

ARTICLE 17. STRATEGIES TO PREVENT AND CONTROL POLLUTION OF GROUNDWATER

A rather similar Article to the last, but describing groundwater rather than surface water.

Article 18. Commission report

In 2015 and every six years thereafter the European Commission shall publish a report with:

- progress of the implementation of this Directive
- status of surface water and groundwater in the Community
- survey of river basin (catchment) management plans, together with recommendations for their improvement
- a summary of each report or recommendation made by Member States
- a summary of any proposals, control measures and strategies made under Article 16
- a summary of responses to comments on earlier implementation reports (by the European Parliament and Council).

The Commission shall convene conferences of interested parties. *[The list of these is interesting but not explained. Competent authorities come first, the European Parliament only second, Non-government Organizations (NGOs), which have mostly amateur members, and few international experts, are before academics, who do most of the research.]*

Article 19. Plans for future Community measures

Annually, the Commission shall present indicative plans affecting water legislation. The Commission will review the Directive in 2019 at the latest, and will propose any necessary amendments. *[Note the "shall" in the first sentence. This is a command. And note the "will" in the second one. This is the expectation.]*

Article 20. Technical adaptations to the Directive

(Various annexes may be adapted to scientific and technical progress, though being kept in line with the relevant Articles.)

Article 21. Committee procedure

(About the committee to assist the Commission.)

Articles 22. Repeals and transitional provisions

This repeals earlier Directives on water quality and monitoring and allows new water quality standards.

Article 23. Penalties

Member States shall determine penalties, which are effective, proportionate and dissuasive, for breaches of Directive standards.

ARTICLE 24. IMPLEMENTATION

Member States shall bring their own laws and administrative provisions by the end of 2003.

ARTICLE 25. ENTRY INTO FORCE

2000.

ARTICLE 26. ADDRESSEES

This Directive is addressed to the Member States.

River Soar, Loughborough.

See Table below: "Four out of nine water companies are falling short of expected environmental standards, according to figures published [today] by the Environment Agency [©] which show that pollution incidents from water company sewerage and clean water assets rose from 1,863 in 2018 to 2,204 in 2019 – the highest number for five years." Jamie Carpenter, Ends Report, October 2020.

Environment Agency Environmental Performance Assessment (EPA) results 2019 for water and sewerage companies

Metric and units	Pollution incidents (sewerage) - per 10,000km of sewer	Serious pollution incidents (sewerage) - per 10,000km of sewer	Discharge permit compliance (STW & WTW) - percentage	Self-reporting of pollution incidents - percentage	National Environment Programme - percentage of plan delivered	Security of Supply Index (SoSI) - score	Performance star rating (out of 4)
Red, amber, green, thresholds	≥50 red >25 amber ≤25 green	≥1.5 red >0.5 amber ≤0.5 green	≤97 red <99 amber ≥99 green	≤55 red <75 amber ≥75 green	≤97 red >97 amber ≥99 green	100 green <100 but ≥99 amber <99 red	
Anglian Water	35 ↓↓	1.6 ↓↓	98.6 ↑	71 ↑	100 ↔	99 ↓↓	**
Northumbrian Water[1]	15 ↓	0.7 ↓↓	96.6 ↓↓↓	80 ↓	100 ↔	100 ↓↔	**
Severn Trent Water	26 ↑	0.3 ↑↑	99.6 ↑↑	78 ↓	99.2 ↔	100 ↔	****
Southern Water	110 ↓↔	1.8 ↔↔	98.8 ↓↓	87 ↑	96.0 ↓↓	100 ↑↑↑	*
South West Water	105 ↓	0.6 ↑	98.7 ↔↔	76 ↓	97.4 ↓↓	100 ↔↔	**
Thames Water	30 ↓	1.4 ↓	99.7 ↑	78 ↓	99.3 ↓	100 ↑↑	***
United Utilities	28 ↓↓	0 ↑	98.5 ↓	90 ↑	99.7 ↑↑	100 ↔↔	***
Wessex Water	22 ↑	0.3 ↑↑	98.5 ↓↓	85 ↑↑	100 ↔↔	100 ↔↔	****
Yorkshire Water	35 ↑	1.3 ↑↑	97.5 ↔↔	73 ↔↔	99.2 ↓	100 ↔↔	***
Sector	37 ↓	0.9 ↔↔	98.7 ↑	80 ↑	98.4 ↓	99.9 ↑	

Metric status	Performance description
Red	Performance significantly below target
Amber	Performance close to or slightly below the target
Green	Performance better than target

Performance star rating	Star rating description
****	Industry leading company 5 or more green metrics and no red metrics
***	Good company 1 or more green metrics and no red metrics
**	Company requires improvement 1 or 2 red metrics and/or zero green metrics
*	Poor performing company more than 2 red metrics

Performance change	Change compared to last year
↑	Improving within class
↑↑	Improved a class
↑↑↑	Improved by 2 classes, e.g. from red to green
↔↔	About the same
↓	Deteriorating within class
↓↓	Deteriorated a class
↓↓↓	Deteriorated 2 classes, e.g. from green to red

[1] Northumbrian Water had 6 discharge permit compliance metric failures. Three of these were due to sample shortfalls associated with laboratory analysis not meeting quality control standards and are not associated with any known environmental impact

Tributary to Great Ouse, Histon, Cambs.

River Avon, Batheaston.

River Stiffkey, Little Walsingham

ANNEXES

Not necessary for this book.

ANNEX 2. DESCRIPTION OF WATER BODY TYPES

SURFACE WATERS

(The brackish transitional water bodies, and the seawater coastal waters are included in the Directive, but omitted from this book, both here and below. See the Directive for details.)

[These appendices are part of the Directive, that is, they are law, an optional extra which can be interpreted variously. Much of the information needed for the proper protection, enhancement and restoration of water bodies is not available: many methods of survey and assessment needed in 2000 and indeed still needed, require further research before being established. But the annexes do show where this work, in the first instance, is necessary. The scope of, in particular, Annexes 2, 5, 6 and 7 is huge and deserves much careful consideration.

We are particularly glad to see plants are considered important, and indeed are placed before animals, as in the traditional English phrase "flora and fauna". Although the nineteenth and early twentieth centuries saw great interest in flora, a lot was by amateurs, and with changing fashions, this has—at least in Britain (and Malta)—mostly lapsed, and been replaced by an amateur interest in birds. The many botanists who used to record every plant species they saw have been largely replaced by yet more people recording—and often travelling far to find— every bird species they can see. Vegetation, however little currently regarded, is the cornerstone of life. Plants can live as a general principle on water, light and minerals from the soil. Animals, equally as a general principle, feed on plants (carnivores eat animals which have fed directly or indirectly on vegetation).

In the nineteenth century, professional workers on waters were of course few, but it was the century of expanding water chemistry; fish and birds were over-exploited and consequently found to be of interest. The twentieth century saw an explosion of professional interest in invertebrate composition and monitoring, particularly in those living on the bottom—the benthic invertebrates. Plants interest fewer and fewer people. There has even been a loss of interest in Britain over the last half century. Wildlife Trusts, for example, usually introduce children to water bodies via pond-dipping, mostly for animals. Plants are usually just "weeds", ignored or considered as background (as in "That is a nice patch of water lilies.").]

Characterisation of surface water body types (*Table 1*)

Each water body is to be identified, and classed as river, lake, artificial or heavily modified. Then it is to be typed by either method A or B. *[Clearly these two methods will, at a later date, be merged into one, once it is seen which works best over what countries and regions.]* System A starts with geographical eco-regions. System B, with types of the "obligatory" and "optional" characters that describe water body types.

TABLE 1. Eco-regions and surface water body types

Rivers System A	
Fixed typology	**Descriptors**
Type	Altitude typology high: > 800m mid-altitude: 200m to 800m lowland: < 200m Size typology based on catchment area small: 10 to 100km² medium: > 100 to 1,000km² large: > 1,000 to 10,000km² very large: > 10,000km² Geology calcareous siliceous organic

Rivers System B	
Alternative characterisation	Physical and chemical factors that determine the characteristics of the river or part of the river and hence the biological population structure and composition
Obligatory factors	altitude latitude longitude geology size
Optional factors	distance from river source energy of flow (function of flow and slope) mean water width mean water slope form and shape of main river bed river discharge (flow) category valley shape transport of solids acid neutralising capacity mean substratum composition chloride air temperature range mean air temperature precipitation

Lake System A	
Eco-region	Eco-region
Type	Altitude typology high: > 800m mid-altitude: 200m to 800m lowland: < 200m Depth typology based on mean depth < 3m 3 to 15m > 15m Size typology based on surface area 0.5 to 1km² 1 to 10km² 10 to 100km² Geology calcareous siliceous organic

Table 1 Continued:/

Lake System B	
Alternative characterisation	Physical and chemical factors that determine the characteristics of the lake and hence the biological population structure and composition
Obligatory factors	altitude latitude longitude depth geology, size
Optional factors	mean water depth lake shape residence time mean air temperature air temperature range mixing characteristics (e.g., monomictic, dimictic, polymictic) acid neutralising capacity background nutrient status mean substratum composition water level fluctuation

ESTABLISHMENT OF TYPE-SPECIFIC REFERENCE CONDITIONS FOR DEFINITIONS OF ECOLOGICAL STATUS CLASSIFICATION

General definition for rivers

This is a general definition of ecological quality. For the purposes of classification the values for the quality features of ecological status for each surface water category are those given in these tables.

Feature—General	
High status	There are no, or only very minor, anthropogenic [man-made] alterations to the values of the physico-chemical and hydro-morphological quality features for the surface water body type from those normally associated with that type under undisturbed conditions. [If, of course, anything approaching that can be found!] The values of the biological quality features for the surface water body reflect those normally associated with that type under undisturbed conditions, and show no, or only very minor, evidence of distortion. These are the type-specific conditions and communities
Good status	The values of the biological quality features for the surface water body type show low levels of distortion resulting from human activity, but deviate only slightly from those normally associated with the surface water body type under undisturbed conditions.
Moderate status	The values of the biological quality features for the surface water body type deviate moderately from those normally associated with the surface water body type under undisturbed conditions. The values show moderate signs of distortion resulting from human activity and are significantly more disturbed than under conditions of good status.

Waters achieving a status below moderate shall be classified as poor or bad.

Waters showing evidence of major alterations to the values of the biological quality features for the surface water body type and in which the relevant biological communities deviate substantially from those normally associated with the surface water body type under undisturbed conditions, shall be classified as poor.

Waters showing evidence of severe alterations to the values of the biological quality features for the surface water body type and in which large portions of the relevant biological communities normally associated with the surface water body type under undisturbed conditions are absent, shall be classified as bad. [Note the importance given to the biology, the flora and fauna.]

DEFINITIONS FOR HIGH, GOOD AND MODERATE ECOLOGICAL STATUS IN RIVERS

BIOLOGICAL QUALITY FEATURES

Feature—Phytoplankton	
High status	The taxonomic composition of phytoplankton corresponds totally or nearly totally to undisturbed conditions. The average phytoplankton abundance is wholly consistent with the type-specific physico-chemical conditions and is not such as to significantly alter the type-specific transparency conditions. Planktonic blooms occur at a frequency and intensity which is consistent with the type-specific physico-chemical conditions.

21

Table 1 Continued:/

Good status	There are slight changes in the composition and abundance of planktonic taxa compared with the type-specific communities. Such changes do not indicate any accelerated growth of algae resulting in undesirable disturbances to the balance of organisms present in the water body or to the physico-chemical quality of the water or sediment. A slight increase in the frequency and intensity of the type-specific planktonic blooms may occur.
Moderate status	The composition of planktonic taxa differs moderately from the type-specific communities. Abundance is moderately disturbed and may be such as to produce a significant undesirable disturbance in the values of other biological and physico-chemical quality features. A moderate increase in the frequency and intensity of planktonic blooms may occur. Persistent blooms may occur during summer months.

Feature—*Macrophytes and phytobenthos*	
High status	The taxonomic composition corresponds totally or nearly totally to undisturbed conditions. There are no detectable changes in the average macrophytic and the average phytobenthic abundance.
Good status	There are slight changes in the composition and abundance of macrophytic and phytobenthic taxa compared with the type-specific communities. Such changes do not indicate any accelerated growth of phytobenthos or higher forms of plant life resulting in undesirable disturbances to the balance of organisms present in the water body or to the physico-chemical quality of the water or sediment. The phytobenthic community is not adversely affected by bacterial tufts and coats present due to anthropogenic activity
Moderate status	The composition of macrophytic and phytobenthic taxa differs moderately from the type-specific community and is significantly more distorted than at good status. Moderate changes in the average macrophytic and the average phytobenthic abundance are evident. The phytobenthic community may be interfered with and, in some areas, displaced by bacterial tufts and coats present as a result of anthropogenic activities.

Feature—*Benthic invertebrate fauna*	
High status	The taxonomic composition and abundance correspond totally or nearly totally to undisturbed conditions. The ratio of disturbance sensitive taxa to insensitive taxa shows no signs of alteration from undisturbed levels. The level of diversity of invertebrate taxa shows no sign of alteration from undisturbed levels.
Good status	There are slight changes in the composition and abundance of invertebrate taxa from the type-specific communities. The ratio of disturbance-sensitive taxa to insensitive taxa shows slight alteration from type-specific levels. The level of diversity of invertebrate taxa shows slight signs of alteration from type-specific levels.
Moderate status	The composition and abundance of invertebrate taxa differ moderately from the type-specific communities. Major taxonomic groups of the type-specific community are absent. The ratio of disturbance-sensitive taxa to insensitive taxa, and the level of diversity, are substantially lower than the type-specific level and significantly lower than for good status.

Feature—*Fish fauna*	
High status	Species composition and abundance correspond totally or nearly totally to undisturbed conditions. All the type-specific disturbance-sensitive species are present. The age structures of the fish communities show little sign of anthropogenic disturbance and are not indicative of a failure in the reproduction or development of any particular species.
Good status	There are slight changes in species composition and abundance from the type-specific communities attributable to anthropogenic impacts on physico-chemical and hydromorphological quality features. The age structures of the fish communities show signs of disturbance attributable to anthropogenic impacts on physico-chemical or hydromorphological quality features, and, in a few instances, are indicative of a failure in the reproduction or development of a particular species, to the extent that some age classes may be missing.
Moderate status	The composition and abundance of fish species differ moderately from the type-specific communities attributable to anthropogenic impacts on physico-chemical or hydromorphological quality features. The age structure of the fish communities shows major signs of anthropogenic disturbance, to the extent that a moderate proportion of the type-specific species are absent or of very low abundance.

Table 1 Continued:/

Feature—Hydrological regime	
High status	The quantity and dynamics of flow, and the resultant connection to groundwaters, reflect totally, or nearly totally, undisturbed conditions.
Good status	Conditions consistent with the achievement of the values specified above for the biological quality features.
Moderate status	Conditions consistent with the achievement of the values specified above for the biological quality features

Feature—River continuity	
High status	The continuity of the river is not disturbed by anthropogenic activities and allows undisturbed migration of aquatic organisms and sediment transport
Good status	Conditions consistent with the achievement of the values specified above for the biological quality features.
Moderate status	Conditions consistent with the achievement of the values specified above for the biological quality features

Feature—Morphological conditions	
High status	Channel patterns, width and depth variations, flow velocities, substrate conditions and both the structure and condition of the riparian zones correspond totally or nearly totally to undisturbed conditions.
Good status	Conditions consistent with the achievement of the values specified above for the biological quality features.
Moderate status	Conditions consistent with the achievement of the values specified above for the biological quality features

Feature—General conditions	
High status	The values of the physico-chemical features correspond totally or nearly totally to undisturbed conditions. Nutrient concentrations remain within the range normally associated with undisturbed conditions.
Good status	Levels of salinity, pH, oxygen balance, acid neutralising capacity and temperature do not show signs of anthropogenic disturbance and remain within the range normally associated with undisturbed conditions. Temperature, oxygen balance, pH, acid neutralising capacity and salinity do not reach levels outside the range established so as to ensure the functioning of the type-specific ecosystem and the achievement of the values specified above for the biological quality features.
Moderate status	Nutrient concentrations do not exceed the levels established so as to ensure the functioning of the ecosystem and the achievement of the values specified above for the biological quality features. Conditions consistent with the achievement of the values specified above for the biological quality features.

Feature—Specific synthetic pollutants	
High status	Concentrations close to zero and at least below the limits of detection of the most advanced analytical techniques in general use.
Good status	Concentrations not in excess of the standards set in accordance with the procedure detailed in section 1.2.6 without prejudice to Directive 91/414/EC and Directive 98/8/EC (< environmental quality standard—EQS).
Moderate status	Conditions consistent with the achievement of the values specified above for the biological quality features.

Feature—Specific non-synthetic pollutants	
High status	Concentrations remain within the range normally associated with undisturbed conditions (background levels —bgl).
Good status	Concentrations not in excess of the standards set in accordance with the procedure detailed in section 1.2.6 without prejudice to Directive 91/414/EC and Directive 98/8/EC (< EQS).
Moderate status	Conditions consistent with the achievement of the values specified above for the biological quality features.

Feature—Phytoplankton	
High status	The taxonomic composition and abundance of phytoplankton correspond totally or nearly totally to undisturbed conditions. The average phytoplankton biomass is consistent with the type-specific physico-chemical conditions and is not such as to significantly alter the type-specific transparency conditions. Planktonic blooms occur at a frequency and intensity which is consistent with the type-specific physico-chemical conditions.
Good status	There are slight changes in the composition and abundance of planktonic taxa compared with the type-specific communities. Such changes do not indicate any accelerated growth of algae resulting in undesirable disturbance to the balance of organisms present in the water body or to the physico-chemical quality of the

Table 1 Continued:/

	water or sediment. A slight increase in the frequency and intensity of the type-specific planktonic blooms may occur.
Moderate status	The composition and abundance of planktonic taxa differ moderately from the type-specific communities. Biomass is moderately disturbed and may be such as to produce a significant undesirable disturbance in the condition of other biological quality features and the physico-chemical quality of the water or sediment. A moderate increase in the frequency and intensity of planktonic blooms may occur. Persistent blooms may occur during summer months.

Feature—Macrophytes and phytobenthos

High status	The taxonomic composition corresponds totally or nearly totally to undisturbed conditions. There are no detectable changes int eh average macrophytic and the average phytobenthic abundance.
Good status	There are slight changes in the composition and abundance of macrophytic and phytobenthic taxa compared with the type-specific communities. Such changes do not indicate any accelerated growth of phytobenthos or higher forms of plant life resulting in undesirable disturbance to the balance of organisms present in the water body or to the physico-chemical quality of the water. The phytobenthic community is not adversely affected by bacterial tufts and coats present due to anthropogenic activity.
Moderate status	The composition of macrophytic and phytobenthic taxa differ moderately from the type-specific communities and are significantly more distorted than those observed at good quality. Moderate changes in the average macrophytic and the average phytobenthic abundance are evident. The phytobenthic community may be interned with, and, in some areas, displaced by bacterial tufts and coats present as a result of anthropogenic activities.

Feature—Benthic invertebrate fauna

High status	The taxonomic composition and abundance correspond totally or nearly totally to the undisturbed conditions. The ratio of disturbance-sensitive taxa to insensitive taxa shows no signs of alteration from undisturbed levels. The level of diversity of invertebrate taxa shows no sign of alteration from undisturbed levels.
Good status	There are slight changes in the composition and abundance of invertebrate taxa compared with the type-specific communities. The ratio of disturbance-sensitive taxa to insensitive taxa shows slight signs of alteration from type-specific levels. The level of diversity of invertebrate taxa shows slight signs of alteration from type-specific levels.
Moderate status	The composition and abundance of invertebrate taxa differ moderately from the type-specific conditions. Major taxonomic groups of the type-specific community are absent. The ratio of disturbance-sensitive to insensitive taxa and the level of diversity, are substantially lower than the type-specific level and significantly lower than for good status.

Feature—Fish fauna

High status	Species composition and abundance correspond totally or nearly totally to undisturbed conditions. All the type-specific sensitive species are present. The age structures of the fish communities show little sign of anthropogenic disturbance and are not indicative of a failure in the reproduction or development of a particular species.
Good status	There are slight changes in species composition and abundance from the type-specific communities attributable to anthropogenic impacts nonphysico-chemical or hydromorphological quality features. The age structures of the fish communities show signs of disturbance attributable to anthropogenic impact on physico-chemical or hydromorphological quality features, and, in a few instance, are indicative of a failure in the reproduction or development of a particular species, to the extent that some age classes may be missing.
Moderate status	The composition and abundance of fish species differ moderately from the type-specific communities attributable to anthropogenic impacts on physico-chemical or hydromorphological quality features. The age structure of the fish communities shows major signs of disturbance, attributable to anthropogenic impacts on physico-chemical or hydromorphological quality features, to the extent that a moderate proportion of the type-specific species are absent or of very low abundance.

HYDROMORPHOLOGICAL QUALITY FEATURES

Feature—Hydrological regime

High status	The quantity and dynamics of flow, level, residence time, and the resultant connection to groundwaters, reflect totally or nearly totally undisturbed conditions.
Good status	Conditions consistent with the achievement of the values specified above for the biological quality features.
Moderate status	Conditions consistent with the achievement of the values specified above for the biological quality features.

Feature—Morphological conditions

High status	Lake depth variation, quantity and structure of the substate, and both the structure and condition of the lake shore zone correspond totally or nearly totally to undisturbed conditions.

Table 1 Continued:/

Good status	Conditions consistent with the achievement of the values specified above for the biological quality features.
Moderate status	Conditions consistent with the achievement of the values specified above for the biological quality features.

PHYSICO-CHEMICAL QUALITY FEATURES

Feature—General conditions	
High status	The values of physico-chemical features correspond totally or nearly totally to undisturbed conditions. Nutrient concentrations remain within the range normally associated with undisturbed conditions. Levels of salinity pH, oxygen balance, acid neutralising capacity, transparency and temperature do not show signs of anthropogenic disturbance and remain within the range normally associated with undisturbed conditions.
Good status	Temperature, oxygen balance, pH, acid neutralising capacity, transparency and salinity do not reach levels outside the range established so as to ensure the functioning of the ecosystem an the achievement of the values specified above for the biological quality features. Nutrient concentrations do not exceed the levels established so as to ensure the functioning of the ecosystem and the achievement of the values specified above for the biological quality features.
Moderate status	Conditions consistent with the achievement of the values specified above for the biological quality features.

Feature—Specific synthetic pollutants	
High status	Concentrations close to zero and at least below the limits of detection of the most advanced analytical techniques in general use.
Good status	Concentrations not in excess of the standards set I accordance with the procedure detailed in section 1.2.6 without prejudice to Directive 91/414/EC and Directive 98/8/EC. (< EQS)
Moderate status	Conditions consistent with the achievement of the values specified above for the biological quality features.

Feature—non-synthetic pollutants	
High status	Concentrations remain within the range normally associated with undisturbed conditions (bgl).
Good status	Concentrations not in excess of the standards set in accordance with the procedure detailed in section 1.2.6 (EQS >bgl). Without prejudice to Directive 91/414/EC and Directive 98/8/EC. (< EQS)
Moderate status	Conditions consistent with the achievement of the values specified above for the biological quality features.

DEFINITIONS FOR MAXIMUM, GOOD AND MODERATE ECOLOGICAL POTENTIAL FOR HEAVILY MODIFIED OR ARTIFICIAL WATER BODIES

Feature—Biological quality features	
Maximum ecological potential	The values of the relevant biological quality features reflect, as far as possible, those associated with the closest comparable surface water body type, given the physical conditions which result from the artificial or heavily modified characteristics of the water body.
Good ecological potential	There are slight changes in the values of the relevant biological quality features as compared with the values found at maximum ecological potential.
Moderate ecological potential	There are moderate changes in the values of the relevant biological quality features as compared with the values found at maximum ecological potential. These values are significantly more distorted than those found under good quality.

Feature—Hydromorphological features	
Maximum ecological potential	The hydromorphological conditions are consistent with the only impacts on the surface water body being those resulting from the artificial or heavily modified characteristics of the water body once all mitigation measures have been taken to ensure the best approximation to ecological continuum, in particular with respect to migration of fauna and appropriate spawning and breeding grounds.
Good ecological potential	Conditions consistent with the achievement of the values specified above for the biological quality features.
Moderate ecological potential	Conditions consistent with the achievement of the values specified above for the biological quality features.

PHYSICO-CHEMICAL QUALITY FEATURES

Feature—General conditions	
Maximum ecological potential	Physico-chemical features correspond totally or nearly totally to the undisturbed conditions associated with the surface water body type most closely comparable to the artificial or heavily modified body concerned. Nutrient concentrations remain within the range normally associated with such undisturbed conditions. The levels of temperature, oxygen balance and pH are consistent with those found in the most closely

Table 1 Continued:/

Good ecological potential	comparable surface water body types under undisturbed conditions. The values for physico-chemical features are within the ranges established so as to ensure the functioning of the ecosystem and the achievement of the values specified above for the biological quality features. Temperature and pH do not reach levels outside the ranges established so as to ensure the functioning of the ecosystem and the achievement of the values specified above for the biological quality features. Nutrient concentrations do not exceed the levels established so as to ensure the functioning of the ecosystem and the achievement of the values specified above for the biological quality features.
Moderate ecological potential	Conditions consistent with the achievement of the values specified above for the biological quality features.
Feature—Specific synthetic pollutants	
Maximum ecological potential	Concentrations close to zero and at least below the limits of detection of the most advanced analytical techniques in general use.
Good ecological potential	Concentrations not in excess of the standards set in accordance with the procedure detailed in section 1.2.6 without prejudice to Directive 91/414/EC and Directive 98/8/EC. (< EQS)
Moderate ecological potential	Conditions consistent with the achievement of the values specified above for the biological quality features.
Feature—Non-specific synthetic pollutants	
Maximum ecological potential	Concentrations remain within the range normally associated with the undisturbed conditions found in the surface water body type most closely comparable with the artificial or heavily modified body concerned (bgl).
Good ecological potential	Concentrations not in excess of the standards set in accordance with the procedure detailed in section 1.2.6* without prejudice to Directive 91/414/EC and Directive 98/8/EC. (< EQS)
Moderate ecological potential	Conditions consistent with the achievement of the values specified above for the biological quality features. *Application of the standards derived under this protocol shall not require reduction of pollutant concentrations below bgl.

Surface water body types

There are different types of water body *[Good!]*. These shall be identified and classified (see Annex 5). Using the biological criteria, high ecological status can be defined. Therefore there can be reference standards, so that any water body can be assessed against the reference high status for that particular water body type.

[A Note from Sylvia Haslam: This is of exceptional interest. For years, indeed, decades, I have been trying to explain to aquatic biologists that the many different kinds of river differ in the plant—and animal—communities they bear, that these communities may be better or worse examples of their type, and that their rating should be against the best examples of that type. Now this principle is enshrined in law—I am pleased!]

- In much-modified water bodies, etc., the rating should be against that which is the higher ecological potential of that type of water body.
- The reference status may be that seen (spatial), that modelled, a mixture of the two, or that by using expert judgement (where pollutant concentrations are concerned, the method of analysis available must be considered.)
- Spatially-based biological reference conditions must be based on enough high-status sites so the conditions set are reliable and reproducible. *[Biological systems are variable!]*

- Modelled reference conditions shall use historic, palaeoecological and other data, and likewise must provide confidence in their consistency and validity.

[Unluckily, few of those in the field are competent in both understanding the ecology and autecology of what they are recording and modelling. Too many models are ecologically flawed. This, of course, is probably just teething troubles.]

- Where variability is so great—and its causes not established—sites should be omitted, but reports should include this and the reasons why.

Identification of pressures

The pressures of influence of human activities on each water body type shall be listed, particularly:

(a) *Pollution*: both point and diffuse and source, pollutions, from urban, industrial, agricultural and other installations and activities. Data shall be collected widely, as specified in the various earlier Directives.

(b) *Abstraction*: both estimation and identification are needed, including seasonal variation (for example, more in the growing or tourist seasons) and annual demand. Loss of water in distribution systems must also be included. (Whilst some unused water returns to "water resources", some is lost to the air or to the sea.)

(c) *Flow regulation*: again both estimation and identification of it and its results are needed. This includes transfers between and diversions from water bodies and overall flow characteristics and water balances.

(d) *Alterations* to the shape, texture, etc. (morphology), of water bodies.

(e) *Other man-made impacts*, their estimation and identification. *[How wise! No general list can include the odd things people may do, e.g., digging a ditch from a garden, throwing old poly-bags out of a canoe, creating an opencast mine, using a new pesticide, planting exotic water lilies.]*

[These impacts are in addition to those included, but not separately identified, in (a) to (d) above: the effects of management. Management is pervasive throughout the lowlands and plains, diminishing but still present in the mountains. Whilst looking at the effect of a factory, it is easy to forget that routine ordinary maintenance is affecting channel shape, water volume, flow, pollution (dilution, silting), flooding, substrate (dredging, cutting, silting), vegetation (cutting, species composition, etc.). All effects on shape, flow and structure are to be found, considered and included.]

(f) *Land use patterns*, estimation and identification of the main urban, industrial and agricultural areas and, where relevant, fisheries and forests.

Assessment of Impact

This is the assessment of the response to these pressures and activities, and whether the water bodies in the catchment basin are likely to meet the objectives set for them by **Article 4 Environmental Objective: protection, enhancement and restoration**. Modelling techniques are allowed.

Any water body considered likely to fail shall have special monitoring and Programmes of Measures.

Edward Henry Holder (1864–1917). "On the Thames near Richmond"

William Mellor (1851–1931). "A View on the Wharfe, Nr. Barden, Yorks"

Richard Wilson (1713–1782). "Westminster Bridge under Construction, 1744"

"The Cathedral and City of Ely, 1810"

GROUNDWATERS

Initial characteristics

This Annex assesses the users of the Directive and how likely it is that environmental objectives may fail. Groundwater bodies may be grouped. Analysis may use existing hydrological, geological, pedological (soil), land use, discharge, abstraction and other data *[a very wide range of speciality subject areas!]*, but shall identify *[Note—shall identify]*:

* the place and boundary of each
* the pressures each is liable to, including: diffuse sources of pollution; point sources; abstraction; artificial recharge
* the general character of the (relevant) rock
* which groundwater bodies have dependent surface water or land ecosystems (for example, wetlands fed from groundwater springs).

Further characteristics

A list needed for those at risk, in order that proper measures shall be taken *[Note, once more— shall be taken]*:

* **geological** characteristics
* **hydrological** characteristics (including hydraulic conductivity, porosity and confinement
* characteristics of the **recharge catchment area** (including thickness, porosity, hydraulic conductivity, and absorptive properties of the deposits and soils)
* **stratification** features in the groundwater
* a list of the surface water-bodies which are **dynamically** *[care!]* connected to the groundwaters
* estimates of the **directions**, exchanges and their rates between them
* the long-term average rate of **recharge**, i.e., how much water comes in
* **chemical compositions**, including the effects of human activities.

Review of the impact of human activity on groundwaters

For those water bodies at risk of failing:

* the **whereabouts** of the larger abstraction points
* the annual average **rates of abstraction** from these
* the **chemical composition** of the abstracted water
* the **places** where water is directly discharged into the groundwater
* the **rates of discharge** at such points
* the **chemical composition** of such water.
* *[and a comprehensive catch-all point]* the **land use** in the catchments recharging the groundwater. This includes the pollution from the land, and man-made changes to recharge the groundwater from, for example, diverting run-off, rain, sealing land (e.g., cementing, building), damming, drainage, and artificial recharge.

Review of the impact of changes in groundwater levels

Identifying groundwater bodies which will not rise above low status because of surface issues, water regulation (including flood protection and land drainage), or human development.

Review of the impact of pollution on groundwater quality

Identifying groundwater bodies which will only reach lower status because of pollution (which is too difficult or expensive to clean up).

ANNEX 3. ECONOMIC ANALYSIS

This shall be detailed enough to give long-term forecasts of supply and demand, estimate of costs, and judgements of the most cost-effective measures.

ANNEX 4. PROTECTED AREAS

These are to be listed and indeed mapped. Their importance is for their value in any of: abstraction, economically significant aquatic species, recreation (including bathing), nutrient-sensitivity or vulnerability *[areas where nitrate fertiliser sinks quickly and greatly to the aquifer]*, protecting habitats or species, and where water status is important in that protection. The summary list shall include the laws or by-laws under which each area is designated.

ANNEX 5. WATER STATUS *[What a comprehensive description of a river this could give!]*

Quality features for the classification of ecological status

MONITORING RIVERS

What are these features?

*   Biological features: composition and abundance of aquatic flora and benthic (bottom) invertebrate fauna; the same for fish, plus their age structure (what proportion are aged 0–1 years, 1–2 years, etc..
*   The hydromorphological features which support these biological features. That is, the hydrological regime (flow, groundwater connections to surface water); the river continuing, and the morphological conditions (depth and width, substrate, shape and structure).
*   The chemical and physico-chemical features which support the biological features. That is, the general components (thermal conditions—temperature, oxygenation, salinity, acidification, and nutrients); and the specific pollutants, Table 2 (priority substances, Table 3, other lists).

Table 2. Indicative list of the main pollutants

1. Organohalogen compounds and substances which may form such compounds in the aquatic environment.
2. Organophosphorous compounds.
3. Organotin compounds.

Table 2 Continued:/

4. Substances and preparations, or the breakdown products of such, which have been proved to possess carcinogenic or mutagenic properties or properties which may affect steroidogenic, thyroid, reproduction or other endocrine-related functions in or via the aquatic environment.

5. Persistent hydrocarbons and persistent and bioaccumulable organic toxic substances.

6. Cyanides.

7. Metals and their compounds.

8. Arsenic and its compounds.

9. Biocides and plant protection products.

10. Materials in suspension.

11. Substances which contribute to eutrophication (in particular, nitrates and phosphates).

12. Substances which have an unfavourable influence on the oxygen balance (and can be measured using parameters such as BOD-Biological Oxygen Demand, COD-Chemical Oxygen Demand, etc.).

Quality features for the classification of surface water status:
- to be investigated for each river, lake, artificial and heavily modified water body.

Normative *[standard or reference]* **definitions of ecological status classifications:**
- for different types of river and lake
- for the maximum, the good and the moderate ecological potential for heavily modified or artificial water bodies
- for the procedure for setting chemical quality standards.

Monitoring of ecological and chemical status for surface waters:
- monitoring designs for surveillance, operations and investigations
- frequency of monitoring
- additional monitoring for protected areas
- standards for monitoring of quality.

Classification and presentation of ecological status:
- comparing biological results
- clarifying ecological status and ecological potential
- classifying chemical status.

Groundwater Status

Groundwater quantitative status:
- classifying quantitative status
- defining quantitative status.

Monitoring of groundwater quantitative status:
- groundwater level monitoring network
- density of monitoring sites
- monitoring frequency
- interpretation and presentation

Table 3. Priority substances in the field of water policy (1)

Name of priority substance	Identified as priority hazardous substance
Alachlor	
Anthracene	(X) (3)
Atrazine	(X) (3)
Benzene	
Brominated diphenylethers (2)	X (4)
Cadmium and its compounds	X
C10-13-chloroalkanes (2)	X
Chlorfenvinphos	
Chlorpyrifos	(X) (3)
1,2-Dichloroethane	
Dichloromethane	
Di(2-ethylhexyl)phthalate (DEHP)	(X) (3)
Diuron	(X) (3)
Endosulfan	(X) (3)
(alpha-endosulfan)	
Fluroanthene (5)	
Hexachlorobenzene	X
Hexachlorobutadiene	X
Hexachlorocyclohexane	X
(gamma-isomer, Lindane)	
Isoproturon	(X) (3)
Lead and its compounds	(X) (3)
Mercury and its compounds	X
Naphthalene	(X) (3)
Nickel and its compounds	
Nonylphenols	X
(4-(para)-nonylphenol)	
Octylphenols	(X) (3)
(para-tert-octylphenol)	
Pentachlorobenzene	X
Pentachlorophenol	(X) (3)
Polyaromatic hydrocarbons	X
(Benzo(a)pyrene),	
(Benzo(b)fluoranthene,	
(Benzo(g,h,i)perylene),	
(Benzo(k)fluoranthene),	
(Indeno(1,2,3-cd)pyrene)	
Simazine	(X) (3)
Tributyltin compounds	X
(Tributyltin-cation)	
Trichlorobenzenes	(X) (3)
(1,2,4-Trichlorobenzene)	
Trichloromethane (Chloroform)	
Trifluralin	(X) (3)

(1)   Where groups of substances have been selected, typical individual representative are listed as indicative parameters (in brackets and without number). The establishment of control will be targeted to these individual substances, without prejudicing the inclusion of other individual representatives, where appropriate.

(2)   These groups of substances normally include a considerable number of individual compounds. At present, appropriate indicative parameters cannot be given.

(3)   This priority substance is subject to a review for identification as possible "priority hazardous substance". The Commission will make a proposal to the European Parliament and Council for its final classification not later than 12 months after adoption of this list. The timetable laid down in Article 16 of the Water Framework Directive for the Commission's proposals of controls is not affected by this review.

(4)   Only Pentabromohiphenylether (CAS-number 32534-81-9).

(5)   Fluoranthene is on the list as an indicator of other, more dangerous Polyaromatic Hydrocarbons.

- surveillance and operational monitoring
- trends in pollutants
- interpretation and presentation of chemical status.

Groundwater chemical status:
- determining status
- defining status.

Monitoring of groundwater chemical status:
- groundwater monitoring network.

Presentation of groundwater status

Lakes

- Biological features: composition, abundance and biomass of phytoplankton; composition and abundance of other aquatic flora; of benthic invertebrate fauna; for fish, the composition, abundance and age structure, etc..
- The hydromorphic features which support the biological features. That is, the hydrological regime (flow, residence time, groundwater connections); and the morphological conditions (depth, substrate, shore structure).
- The chemical and physico-chemical features which support the biological features. That is, general ones (transparency, thermal conditions, oxygenation conditions, salinity, acidification status, nutrients); and the specific pollutants (priority substances, Table 3).

[There are two sections on transitional and coastal waters, which are omitted from this book.]

Artificial and heavily modified surface water bodies

The quality features applied shall be those of river, lake, transitional or coastal water body, whichever the site is closest to.

Procedure for the setting of chemical quality standards by Member States
[In order to carry these out, full laboratory equipment and analyses are needed. Those with these laboratories will be working with the complete Directive, and have no need of such Introduction as is given here.]
The standards shall be subject to peer review and public consultation, including allowing increasingly precise safety factors to be calculated.

Monitoring of ecological status and chemical status for surface waters
Coherent and *comprehensive* are the watchwords for the establishment of monitoring networks and classifications.

River basin catchment shall have a surveillance monitoring plan, an operational monitoring plan and, for some, an investigating monitoring plan. Monitoring programmes shall include each quality feature *[plants as well as animals, both flowering rush (Butomus umbellatus) and water vole (Arvicola terrestris)!]*—whichever taxonomic level (genus, variety, etc.) gives adequate confidence and precision shall be used *[even if tiresome for those monitoring]*.

Design of surveillance monitoring
This is to be able to:
- supplement and validate impact assessments

The beautiful flowering rush (*Butomus umbellatus*) left, and "cuddly" water vole (*Arvicola terrestris*).

- design efficient and effective monitoring programmes
- assess long-term changes
- assess how far these come from widespread human activities.

Catchment management plans can then be based on these plus impact assessments.

There shall be enough monitoring points to assess catchment status. This covers:
- flow rates
- water volume (including larger lakes and reservoirs)
- sites identified under earlier rules (77/795/EEC).

Each site shall be monitored for one year whilst its management plan operates, including:
- all biological quality features
- all hydromorphological quality features
- all general physico-chemical quality features
- priority list pollutants
- significant other pollutants,

unless the water body status is already Good, and is likely to remain so. The monitoring can be decreased (to once in three years).

[Note the value of repetition here: no one can omit diatoms because "why bother!", or, rather, they must know they are being "derelict" if they do omit them.]

Design of operational modelling

The design is to make it possible to:
- establish which water bodies are likely to fail their environmental objective
- assess the effects of the Programmes of Measures on the objectives.

[The difference between this and surveillance monitoring is described above.]

The programme may, as is only reasonable, be amended by new information, particularly to decrease monitoring frequency when impact is removed or found to be insignificant. *[Frequency should not be dropped because of laziness or convenience.]*

Monitoring sites shall be selected because they have:
- risk of failure
- pollution by priority list substances
- significant pollution from one or all point sources
- significant pollution from one or all diffuse sources
- significant hydromorphological pressures.

Quality features indicating man-made pressures shall be monitored as:
- the biological features most sensitive to the pressures;
- all priority (pollutant) substances discharged, and other pollutants discharged in significant quantities;
- the hydromorphological features most sensitive to the pressures.

Design of investigative monitoring
This is to be done to:
- identify unknown harm
- assess the magnitude and impacts of accidental pollution,

and a Programme of Measures shall be established to cover this.

Frequency of monitoring
This shall be in accordance with Table 4, unless greater intervals would be justified in terms of technical knowledge and expert judgement.

Monitoring frequencies shall be selected which take account of the variability in parameters resulting from both natural and anthropogenic conditions. The times at which monitoring is undertaken shall be selected so as to minimise the impact of seasonal variation on the results, and thus ensure that the results reflect changes in the water body as a result of changes due to anthropogenic pressure. Additional monitoring during different seasons of the same year shall be carried out, where necessary, to achieve this objective).

[Note that there are two ways: considered equal. A good piece of machinery, or 20 years' experience, roughly. Think about it. How do you judge whether a printer or a pair of shoes are worth buying? Why? It is worth pondering this. Why, indeed? I can look at an (ordinary) river and say "Organic pollution, Moderate" after 10 seconds or so. The next question, "How ever? Why ever?" is more likely to take 10 minutes, but would lead to the greater precision of a, say, "C" grade. The expert judgement was based on innumerable previous examples studied in much detail, and the general impression—invisible to the others present. The slow method is to class the landscape, land use and rock type, the size category, check that the morphological characters fit these, that the depth is the standard depth for that river type, the same for turbidity, and so on. Then each macrophyte species present is sought for, named and listed together with its abundance and river distribution. Then the habitat preferences of each of these are listed (floating-leaved, nutrient-rich, lime-medium, usually medium depth and moderate flow, in patches of coarser substrate, etc.), other relevant pressures such as walled, children paddling, discoloured plants in summer, pipes with turbid water trickling in, heavy shade downstream. From this list the definitive quiz is answered, the answer given.]

Table 4. Frequency of monitoring

Quality element	Rivers	Lakes
Biological		
Phytoplankton	6 months	6 months
Other aquatic flora	3 years	3 years
Macro-invertebrates	3 years	3 years
Fish	3 years	3 years
Hydromorphological		
Continuity	6 years	
Hydrology	continuous	1 month
Morphology	6 years	6 years
Physico-chemical		
Thermal conditions	3 months	3 months
Oxygenation	3 months	3 months
Salinity	3 months	3 months
Nutrient status	3 months	3 months
Acidification status	3 months	3 months
Other pollutants	3 months	3 months
Priority substances	1 month	1 month

Surveillance monitoring shall be carried out at least once in every monitoring period. Operational monitoring shall be done as in the Table above. (The timing must fit in with seasonal changes, so that anthropogenic pressures can be detected.)

Additional monitoring requirements for protected areas

Supplementary monitoring is needed for:

- abstracted water for larger populations (12 times a year for over 30,000 people, 8 times for 10,000–30,000, though only 4 for under 10,000)
- habitat and species protection areas (two types: failing sites, and sites where assessment of the effectiveness of programmes of measures are needed).

Standards for monitoring or quality standards

There are half a dozen earlier standards for (mainly benthic) invertebrates. For other groups, standards that are yet to be established shall be followed.

Classification and presentation of ecological status
The five grades to be used are High, Good, Moderate, Poor and Bad. How to establish these, establish the boundaries between them, and how to make monitoring systems comparable is a highly technical matter, and is described accordingly.

Presentation of monitoring results and classification of ecological status and ecological potential
Classification shall be by whichever is lower of biological and physico-chemical status. On the charts, rivers shall be coloured *[just like Primary School!]*: High = blue *[sky and sea]*; Good = green; Moderate = yellow; Poor = orange *[warning! Traffic lights are about to change]*; Bad = red *[red for danger, warning and stop lights on, e.g. roads, train tracks].*

Artificial and heavily modified water bodies are, in addition, to be marked with stripes for their ecological potential; light grey for artificial, dark grey for heavily modified habitats; both striped with the colour given for the more natural water body *[except that blue is not used. Green is Good and above.]*

Presentation of monitoring results and classification of chemical status
A water body complying with all environmental quality standards is classed as Good, and coloured blue on charts. One that does not comply is "failing to achieve Good", and coloured red.

GROUNDWATER

Groundwater Quantitative Status

Lambeth Riverside pre-1860 (before the Thames Flood-protecting Albert Embankment was built)

Parameters for the classification of quantitative status
Groundwater Level Regime.
Definition of quantitative status

Good status is when the groundwater level is not lowered by long-term annual average abstraction *[where the quantity of groundwater stays the same]*. Environmental objectives are achieved, associated surface water status is not decreased and neither are land ecosystems depending on the groundwater. Any man-made changes in level do not cause sea water intrusion, nor is any flow change likely to lead to sea water intrusion.

Monitoring of groundwater quantitative status
Groundwater level monitoring network
This shall provide a reliable assessment of status, including available resources. It should be mapped.

Density of monitoring sites
There shall be enough sites to assess in general, and in particular those bodies liable to fail their environmental objectives.

Monitoring frequency
This shall be enough to assess in general the status of all bodies, and in particular those bodies liable to fail their environmental objectives.

Interpretation and presentation of groundwater quantitative status
The assessments from the network shall be mapped, with good status coloured green, poor status red.

Groundwater chemical status
Parameters for the determination of groundwater chemical status
Conductivity and concentration of pollutants. *[Oh dear, what a pity to drop the holistic understanding here. Groundwater, sooner or later, comes out to become surface water, and spring or flush or run-off water becomes stream water and bears life. Life, biological status, reflects the natural chemical status, not just the pollutants. Chalk water bears a different flora and fauna to, say, granite water!]*

Definition of good groundwater chemical status
Pollutant concentrations shall not:
• indicate saline or other intrusion
• exceed the quality standards laid down
• lead to failure in environmental objectives, damage to either groundwater or land ecosystems depending on them. *[At last! But not as comprehensive as would be hoped.]* Likewise conductivity changes may not be enough to indicate saline or other intrusion.

Monitoring of groundwater chemical status
Groundwater monitoring network
This is to provide a coherent and comprehensive overview and detect long-term man-made increases in pollutants.

The assessments (above) lead to management plans, and to surveillance and operational monitoring, which themselves need estimates of their level of confidence and the precision of the results. *[Very wise! It is far too easy to waffle or model without back-checking what is actually in the water body!]*

Surveillance monitoring
This is to:
- supplement and validate impact assessments
- use in assessing long-term trends, natural and man-made
- be enough to identify water bodies at risk.

Core monitoring shall cover: oxygen content, pH, conductivity, nitrate and ammonium, with extra monitoring for water bodies at risk from other parameters.

Operational monitoring
This is to:
- establish chemical status
- use in assessing long-term man-made upwards trends *[in other words: sites getting worse]*
- monitor water bodies at risk of failing their environmental objectives.

It shall be done often enough to detect impacts, and at least once a year.

Identification of trends in pollutants
Identifying increases, and reversing them.

Interpretation and presentation of groundwater chemical status
In assessing status, the aggregated results of monitoring points shall be used and the average result at each point shall be calculated.

The groundwater map shall be coloured green for Good, red for Poor, with black dots for long-term (man-made) pollution increases, and blue dots for pollution decreases. The maps are to be included in catchment management plans.

Presentation of groundwater status
(Where the status maps are difficult.)

ANNEX 6. LISTS OF MEASURES TO BE INCLUDED WITHIN THE PROGRAMMES OF MEASURES

The new Programmes must include the eleven earlier Directives, from Birds (1979) and Habitats (1992) to Bathing Water (1976) and Sewage Sludge (1986). Supplementary measures are required where relevant, and include:

- abstraction controls
- administrative instruments
- aquifer recharge (artificial)
- construction projects
- demand management
- desalination plants
- economic instruments
- educational projects
- emission controls
- environmental agreements
- good practice codes
- laws
- other relevant measures
- rehabilitation projects
- research etc., projects
- water-efficient technology
- wetland restoration, etc.

[A catch-all!]

Annex 7. River Basin Management Plans (RBMPs)

These cover:

General description (Article 4, Annex 2)

For surface waters:
- mapping location and boundary
- mapping ecoregions and surface water body types in the region
- identifying reference conditions for each type.

For groundwaters:
- mapping location and boundary.

A summary of significant pressures and man-made impacts, including:
- estimate of point source pollution
- estimate of diffuse source pollution, with land use summary
- estimate of pressures on quantitative status (including abstraction)
- analysis of other impacts of human activity on the status.

Identification and mapping of protected areas (Article 6, Annex 4)
Mapping monitoring networks (Article 8, Annex 5) and the results of the monitoring for surface water (ecological and chemical), groundwater (chemical and quantitative) and protected areas.

[Here is one of the rare omissions in the directive. Surface water should also be monitored for quantitative status as the lowering of water level and drying up of water bodies is the source of their greatest harm over the past century or so (see Drying Up Book, RFS1, in this series).]

The environmental objectives (Article 4)

Summary economic analysis (Article 5, Annex 3)

Summary programmes of measures (Article 11), together with how these will achieve the objectives [How wise! Not just copied from somewhere else!]:
- summary of protective measures
- summary of measures for cost recovery
- summary of measures for Article 7
- summary of controls of abstraction, etc.
- summary of controls on point source discharges and other activities under Article 11
- summary of measures eliminating or decreasing priority substances (Article 16)
- summary of measures to prevent or reduce the impact of accidental pollutions
- summary of measures for those water bodies likely to fail
- details of supplementary measures needed for the objectives
- measures to avoid increases in sea pollution coming from the land.

A register and summary of any more detailed programmes

A summary of public information and consultation measures, and action therefrom

A list of competent authorities (Annex 1)

Contact procedures, etc., for information in Article 14, of control measures for Article 11, and the monitoring data in Article 8 and Annex 5.

The first update of the river basin management plan and subsequent updates shall also include:
- summary of changes since the earlier management plan
- assessment of progress towards the environmental objectives

41

- summary, and explanation of measures listed earlier which have not reached their objective
- summary of any extra interim measures adopted.

[It is effectively impossible for the first version of any law to be fool-proof, unable to be evaded or avoided, and delivering only all the expected results. However, it is difficult to omit parts here because of laziness, difficulty, experience or presumed lack of money, without it showing in what has been provided. The list is just so complete!]

ANNEX 8. INDICATIVE LIST OF THE MAIN POLLUTANTS
(Presented in Annex 2 above.)

ANNEX 9. EMISSION-LIMIT VALUES AND ENVIRONMENTAL QUALITY STANDARDS

Five earlier Directives apply here: mercury (1982), cadmium (1986), mercury (1984), hexachlorocyclohexane (1984) and dangerous substances (1986). As high-level chemical analysis is involved, it is not discussed here.

ANNEX 10. LIST OF PRIORITY SUBSTANCES IN THE FIELD OF WATER POLICY
(Presented in Annex 2 above.)

ANNEX 11. MAPS
(Omitted.)

River Danube, Hungary

River Sheppey, Croscombe

River Cam, Upware

South West Peak

River Adda, Bangor, Wales

44

ADDENDUM

DIRECTIVE ON THE PROTECTION OF GROUNDWATER AGAINST POLLUTION AND DETERIORATION (SUMMARY)

Directive 2006/118/EC of the European Parliament and of the Council

Whereas:

(1) Groundwater is a valuable natural resource and as such should be protected from deterioration and chemical pollution, This is particularly important for groundwater-dependent ecosystems and for the use of groundwater in water supply for human consumption.

(2) Groundwater is the most sensitive and the largest body of freshwater in the European Union and, in particular, also a main source of public drinking water supplies in many regions.

(3) Groundwater in bodies of water used for the abstraction of drinking water or intended for such future use must be protected in such a way that deterioration in the quality of such bodies of water is avoided in order to reduce the level of purification treatment required in the production of drinking water.

(4) An objective is to achieve water quality without significant impacts on, and risks to, human health and the environment.

(5) In order to protect the environment as a whole, and human health in particular, detrimental concentrations of harmful pollutants in groundwater must be avoided, prevented or reduced.

(6) Measures to prevent and control groundwater pollution should be adopted.

(7) Quality standards and threshold values should be established, and methodologies based on a common approach developed to assess the chemical status of bodies of groundwater.

(8) Quality standards for nitrates, plant protection products and biocides should be set as Community criteria for the assessment of the chemical status of bodies of groundwater.

(9) The protection of groundwater may in some areas require a change in farming or forestry practices, which could entail a loss of income. The Common Agricultural Policy provides funding.

(10) Groundwater chemical status provisions do not apply to high naturally-occurring levels of substances or ions or their indicators, or to temporary changes in flow and composition (which are not regarded as intrusions).

(11) Criteria should be established for the identification of any significant and sustained upward (i.e. worsening) trends in pollutant concentrations.

(12) Member States should, where possible, use statistical procedures, provided they comply with international standards and contribute to the comparability of results of monitoring between Member States over long periods.

(13) (Repeating an earlier Directive.)

(14) The list of Hazardous Substances (above) applies also to groundwater.

(15) Measures to prevent or limit inputs of pollutants into bodies of groundwater used for or intended for future use for the abstraction of water intended for human consumption are needed.

(16) (Frontiers.)

(17) Reliable and comparable methods for groundwater monitoring are important.

(18) Exemptions from limits should be based on transparent criteria and be detailed in the river basin management plans.

(19) Impacts on the level of environmental protection should be analysed.

(20) Research should be conducted in order to provide better criteria for ensuring groundwater ecosystem quality and protection. Where necessary, the findings obtained should be taken into account when implementing or revising this Directive. Such research, as well as dissemination of knowledge, experience and research findings, needs to be encouraged and funded. *[Does it not?!]*

(21) (Transitional measures are needed during the period between the date of implementation of this Directive and the date from which Directive 80/68/EEC is repealed.)

(22) (Various controls and authorisations.)

(23) Supplementary measures Member States may choose to adopt as part of the Programme of Measures include: legislative instruments, administrative instruments, and negotiated agreements for the protection of the environment. (This allows measures better than those specified, if wanted by a Member State.)

ADOPTED DIRECTIVE

THE EUROPEAN PARLIAMENT AND THE COUNCIL OF THE EUROPEAN UNION HAVE ADOPTED THIS DIRECTIVE:

ARTICLE 1. PURPOSE

1. This Directive establishes specific measures as provided for in the Water Framework Directive in order to prevent and control groundwater pollution. These measures include in particular:

   (a) criteria for the assessment of good groundwater chemical status;
   (b) criteria for the identification and reversal of significant and sustained upward trends and for the definition of starting points for trend reversals.

2. This Directive also complements the provisions preventing or limiting inputs of pollutants into groundwater already contained in the Water Framework Directive, and aims to prevent the deterioration of the status of all bodies of groundwater.

Article 2. Definitions

The following definitions shall apply in addition to those laid down in Article 2 above:

1. *groundwater quality standard* means an environmental quality standard expressed as the concentration of a particular pollutant, group of pollutants, or indicator of pollution in groundwater, which should not be exceeded in order to protect human health and the environment;

2. *threshold value* means a groundwater quality standard set by Member States in accordance with Article 3;

3. *significant and sustained upward trend* means any statistically and environmentally significant increase of concentration of a pollutant, group of pollutants, or indicator of pollution in groundwater for which trend reversal is identified as being necessary in accordance with Article 5;

4. *input of pollutants into groundwater* means the direct or indirect introduction of pollutants into groundwater as a result of human activity;

5. *background level* means the concentration of a substance or the value of an indicator in a body of groundwater corresponding to no, or only very minor, anthropogenic alterations to undisturbed conditions;

6. *baseline level* means the average value measured at least during the reference years 2007 and 2008 on the basis of monitoring programmes implemented under Article 8 of the Water Framework Directive, or, in the case of substances identified after these reference years, during the first period for which a representative period of monitoring data is available.

Article 3 Criteria for assessing groundwater chemical status

1. For the purposes of this assessment of the chemical status of a body or a group of bodies of groundwater pursuant to Section 2.3 of Annex V to the Water Framework Directive, Member States shall use the following criteria:

   (a) groundwater quality standards as referred to in Annex I;

   (b) threshold values to be established by Member States in accordance with the procedure set out in Part A of Annex II.

The threshold values applicable to good chemical status shall be based on the protection of the body of groundwater having particular regard to its impact on, and interrelationship with, associated surface waters and directly dependent terrestrial ecosystems and wetlands and shall *inter alia* take into account human toxicology and ecotoxicology knowledge.

2. Threshold values can be established at the national level, at the level of the river basin district or the part of the international river basin district falling within the territory of a Member State, or at the level of a body or a group of bodies of groundwater.

3. Member States shall establish threshold values pursuant to paragraph 1(b) for the first time by 22 December 2008.

All threshold values established shall be published in the river basin management plans to be submitted in accordance with Article 13 of the Water Framework Directive, and including a summary of the information set out in Part C of Annex II to this Directive.

4. Member States shall amend the list of threshold values whenever new information on pollutants, groups of pollutants, or indicators of pollution indicates that a threshold value should be set for an additional substance, that an existing threshold value should be amended, or that a threshold value previously removed from the list should be re-inserted, in order to protect human health and the environment.

Threshold values can be removed from the list when the body of groundwater concerned is no longer at risk from the corresponding pollutants, groups of pollutants, or indicators of pollution.

Any such changes to the list of threshold values shall be reported in the context of the periodic review of the river basin management plans.

5. The Commission shall publish a report by 22 December 2009 on the basis of the information provided by Member States in accordance with paragraph 5.

ARTICLE 4. PROCEDURE FOR ASSESSING GROUNDWATER CHEMICAL STATUS

1. Member States shall use the procedure described in paragraph 2 below to assess the chemical status of a body of groundwater. Where appropriate, Member States may group bodies of groundwater in accordance with Annex V to the Water Framework Directive when carrying out this procedure.

2. A body or a group of bodies of groundwater shall be considered to be of good chemical status when:

   (a) the relevant monitoring demonstrates that the conditions set out in Table 2.3.2. of Annex V to the Water Framework Directive are being met; or
   (b) the values for the groundwater quality standards listed in Annex I and the relevant threshold values established in accordance with Article 3 and Annex II are not exceeded at any monitoring point in that body or group of bodies of groundwater; or
   (c) the value for a groundwater quality standard or threshold value is exceeded at one or more monitoring points but an appropriate investigation in accordance with Annex III confirms that:
      (i) on the basis of the assessment referred to in paragraph 3 of Annex III, the concentrations of pollutants exceeding the groundwater quality standards or threshold values are not considered to present a significant environmental risk, taking into account, where appropriate, the extent of the body of groundwater which is affected;
      (ii) the other conditions for good groundwater chemical status set out in Annex V to the Water Framework Directive are being met, in accordance with paragraph 4 of Annex III to this Directive;
      (iii) for bodies of groundwater identified in accordance with Article 7(a) of the Water Framework Directive, the requirements of Article 7(3) of that Directive are being met, in accordance with paragraph 4 of Annex III to this Directive;
      (iv) the ability of the body of groundwater or of any of the bodies in the group of bodies of groundwater to support human uses has not been significantly impaired by pollution.

3. Choice of the groundwater monitoring sites has to satisfy the requirements of Section 2.4 of Annex V to the Water Framework Directive on being designed so as to provide a coherent

and comprehensive overview of groundwater chemical status and to provide representative monitoring data.

4. Member States shall publish a summary of the assessment of groundwater chemical status in the river basin management plans in accordance with Article 13 of the Water Framework Directive.

   This summary, established at the level of the river basin district or the part of the international river basin district falling within the territory of a Member State, shall also include an explanation as to the manner in which exceedences of groundwater quality standards or threshold values at individual monitoring points have been taken into account in the final assessment.

5. If a body of groundwater is classified as being of Good chemical status in accordance with paragraph 2(c), Member States in accordance with Article 11 of the Water Framework Directive, shall take such measures as may be necessary to protect aquatic ecosystems, terrestrial ecosystems and human uses of groundwater dependent on the part of the body of groundwater represented by the monitoring point or points at which the value for a groundwater quality standard or the threshold value has been exceeded.

ARTICLE 5. IDENTIFICATION OF SIGNIFICANT AND SUSTAINED UPWARD [WORSENING] TRENDS AND THE DEFINITION OF STARTING POINTS FOR TREND REVERSALS

1. Member States shall identify any significant and sustained upward trend and define the starting point for reversing that trend, in accordance with Annex IV.

2. Member States shall reverse trends which present a significant risk of harm to the quality of aquatic ecosystems or terrestrial ecosystems, to human health, or to actual or potential legitimate uses of the water environment, through the Programme of Measures referred to in Article 11 of the Water Framework Directive, in order to reduce pollution and prevent deterioration.

3. Member states shall define the starting point for trend reversal (Annex I, Article 3, Annex IV.)

4. In the river basin management plans to be submitted in accordance with Article 13 of the Directive, Member States shall summarise:
   (a) the way in which the trend assessment has shown that those bodies are subject to a significant and sustained upward trend or a reversal of that trend; and
   (b) the reasons for the starting points defined pursuant to paragraph 3.

5. Plumes of pollution in bodies of groundwater must be studied. The results shall be summarised in the river basin management plans when submitted.

ARTICLE 6. MEASURES TO PREVENT OR LIMIT INPUTS OF POLLUTANTS INTO GROUNDWATER

1. The Programme of Measures established in accordance with Article 11 of that Directive is to include:

(a) all measures necessary to prevent hazardous inputs into groundwater.

(b) for listed but non "hazardous" pollutants considered by Member States to present an existing or potential risk of pollution, so as to ensure that such inputs do not cause deterioration. Measures shall take account, at least, of established best practice.

2. Inputs of pollutants from diffuse sources of pollution having an impact on the groundwater chemical status shall be taken into account whenever technically possible.

3. Without prejudice to any more stringent requirements in other Community legislation, Member States may except from the measures required by paragraph 1 inputs of pollutants that are:

(a) the result of direct discharges authorised by the EC;

(b) considered by the competent authorities to be of a quantity and concentration so small as to obviate any present or future danger of deterioration in the quality of the receiving groundwater; *[Humph! How can this be done?]*

(c) the consequences of accidents or exceptional circumstances of natural cause that could not reasonably have been foreseen, avoided or mitigated;

(d) the result of authorised artificial recharge or augmentation of bodies of groundwater;

(e) in the view of the competent authorities incapable, for technical reasons, of being prevented or limited without using:

    (i) measures that would increase risks to human health or to the quality of the environment as a whole; or

    (ii) disproportionately costly measures to remove quantities of pollutants from, or otherwise control their percolation in, contaminated ground or subsoil *[and who decides what is disproportionate?]*; or

(f) the result of interventions in surface waters for the purposes, amongst others, of mitigating the effects of floods and droughts, and for the management of water and waterways, including at international level. Such activities, including cutting, dredging, relocation and deposition of sediments in surface water, shall be conducted in accordance with general binding rules, and, where applicable, with permits and authorisations on the basis of such rules, developed by Member States for that purpose, provided that such inputs do not compromise the achievement of the environmental objectives established for the water bodies concerned in accordance with Article 4(1)(b) of the Water Framework Directive. *[Yes, but...how much damage can this do?]*

The exemption provided for in points 3(a) to (f) above may be used only where the Member States' competent authorities have established that efficient monitoring of the bodies of groundwater concerned, in accordance with point 2.4.2. of Annex V to the Directive, or other appropriate monitoring, is being carried out. *[Oh Yes?]*

4. The competent authorities of the Member States shall keep an inventory of the exemptions referred to in paragraph 3 for the purpose of notification, upon request, to the Commission.

ARTICLE 7. TRANSITIONAL ARRANGEMENTS

ARTICLE 8. TECHNICAL ADAPTATIONS

1. Parts A and C of Annex II and Annexes III and IV may be amended, in the light of scientific and technical progress.

50

2. Annex II may be amended in order to add new pollutants or indicators.

ARTICLE 9. COMMITTEE PROCEDURE

ARTICLE. 10 REVIEW

Without prejudice to Article 8, the Commission shall review Annexes I and II to this Directive by 16 January 2013, and thereafter every six years. Based on the review, it shall, if appropriate, come forward with legislative proposals.

ARTICLE 11 EVALUATION

The report by the Commission provided for under Article 18(1) of the Directive shall for groundwater include an evaluation of the functioning of this Directive in relation to other relevant environmental legislation, including consistency therewith.

ARTICLE 12 IMPLEMENTATION

Member States shall bring into force the laws, regulations and administrative provisions necessary to comply with this Directive before 16 January 2009. *[Oh dear!]* They shall forthwith inform the Commission thereof.

AFTERWORD

Now, turn again to Table 1, that very long table (pages 20–26). Think of something you would like to know about a river. Does it rise in a hill? See page 20, start of Table 1: "Altitude". What fish are there? See page 22: "Fish fauna". Is it just part of waste water treatment? See page 25: "artificial water bodies". Are unknown pollutants suspected? This obviously cannot be checked without further information, but the last section on page 25 is where you start with this last question. It is extraordinarily comprehensive—and those involved in descriptions might want to enlarge a copy, and even put it on the wall. "Have I considered this one? And that one? Do I need to?..."

Even the best, though, can be superseded by better in a few decades. It will be important, if Britain stays out of the EU, that we take notice of future developments and, in some form, incorporate them into British law.

What of now? As described in the Introduction above, when Britain leaves the EU all laws remain as our laws unless and until parliament changes them. What with the Covid 19 pandemic and the inertia so typical of Britain, it will be a long time—if ever—before parliament takes much interest in what size and shape of pipes in culverts is best for the movement of water shrimps, or *Potamogeton* fruits! So obeying the WFD whilst keeping watch over what parliament might decree (for example, for

Scottish salmon) should keep everyone on the right side of the law. (Remember Statutory Instruments, being legal, can come quickly if scandals arise.)

It is usually easier to say what is wrong, than to say why it went wrong or how to put it right! The excellent descriptions required by the WFD bring up difficulties, for instance, the amalgamation of rivers. In various parts, like W England or Malta, various short rivers are combined and treated as one: so a mainly limestone and a mainly clay river are classed as "single" rivers. There are various rivers which have effectively chalk and clay tributaries joining a main river containing elements of both. This is, though, one river, so what is the difference? Just that it is only the one, separate river.

Different parts of a river may have different influences which have developed differently. It may start clean, but acquire much motorway run-off, farm run-off or town sewage works effluent. Farmland beside the river could be arable (agrichemicals and silt, primarily), grass (less chemicals and silt, but manure and often disturbance of bank and, if shallow, of the bed also), even wood (beneficial if undisturbed, shaded but not polluted). Or Programmes of Measures different for all the innumerable other possibilities, including boat disturbance, weirs, intense fisheries, paddling dogs, fords and more, let alone the attentions of the authorities' dredging which can alter depth, flow, substrate, chemistry, width and bank (height, type of substrate, slope and management).

Regrettably, with loss of EU backing imminent, some authorities have already let oversight slip (for example, collecting but not analysing or publishing the results of monitoring: showing how important the WFD has been in the proper treatment of rivers).

The EU can also take a broader view, seeing features and trends across this very large area which are not noticed, or are ignored, within the confines of one small country.

Blakeney, Norfolk

Rome, Italy

Paxton Pits Nature Reserve, Cambs

53

THE RIVER FRIEND SERIES

This series of small books is designed for people with a general or specific interest in rivers. Please visit the River Friend Website for an up to-date list of PUBLISHED Titles: **http://www.riverfriend.tinasfineart.uk**

Standalone* Titles in the Series include:

A PROLOGUE TO THE SERIES: Plant identification and Glossary of Terms (ISBN 978 1 9162096 2 6)
DRYING UP (ISBN 978 1 9162096 1 9)
STREAM STORY I: A Riveting Riverscape—River Brue, Somerset (ISBN 978 1 9162096 0 2)
INTERPRET: What do Plants Tell us? (ISBN 978 1 9162096 5 7)
REED—ON THE EDGE (ISBN 978 1 9162096 4 0)
An Introduction to the WATER FRAMEWORK DIRECTIVE (ISBN 978 1 9162096 3 3)
WATER: Clean and Dirty (ISBN 978 1 9162096 7 1)
STREAM STORY: A Brook in Transit: Bourn Brook, Cambridge (ISBN 978 1 9162096 8 8)
Vegetation Changes Over Time. Is there FREEZE FRAME? (ISBN 978 1 9162096 6 4)
CHANGE: What a Disaster! (ISBN 978 1 9162096 9 5)
LOOK AT THE BOTTOM
How to lose Fresh Water in Under Two Centuries. The Example of MALTA
VEGETATION PATTERNS
IN THE WATER
THE WATERS OF WELLS
RESTORE, REHABILITATE, IMPROVE
AWFUL ALIENS
WHAT RIVERS DO FOR US
STREAM STORY: Another Riveting Riverscape—River Cam, Cambridge

*Each book is about a different subject so the series can be read in any order

About the Authors

Sylvia Haslam is a botanist and river culture, etc., specialist. Anyone wanting to find out more should look at the publications list on her website (http://www.riversandreeds.co.uk). Her publications specific to this series are listed in the book entitled *A PROLOGUE TO THE SERIES: Plant identification and Glossary of Terms.*

Tina Bone has worked as a self-employed Desktop Publisher for many years until she changed career to work as a Professional Artist from March 2005. To view Tina's resumé and artwork please visit her website: http://www.tinasfineart.uk.

Lightning Source UK Ltd.
Milton Keynes UK
UKHW020954190821
389105UK00007B/179